ORGAN MUSIC
FOR MANUALS ONLY

33 Works
by Berlioz, Bizet, Franck,
Saint-Saëns and Others

Selected and with an Introduction by
Rollin Smith

DOVER PUBLICATIONS, INC.
Mineola, New York

Bibliographical Note

This Dover edition, first published in 2001, is a new compilation of works originally published separately in authoritative early editions. We are indebted to organist/author Rollin Smith for providing the scores and an introduction prepared specially for this edition.

International Standard Book Number
ISBN-13: 978-0-486-41887-2
ISBN-10: 0-486-41887-1

Manufactured in the United States by RR Donnelley
41887106   2015
www.doverpublications.com

Contents

NOTES ON THE MUSIC

Organ music implies not only one or more keyboards but also a pedalboard. Yet, there is a significant body of literature by important composers written on two staves and intended for manuals alone. Some was conceived for organ pipes played by a clockwork, some for the nineteenth-century free-reed instrument, the harmonium or reed organ, and other works as either/or combinations of piano or harmonium, or organ or harmonium; much was written for organists of small churches who may not have had sufficient technique to play with their feet as well as with their hands, while other works were intended for performance in the home.

Organ registrations not indicated in the scores are left to the player's discretion. However, those works that include harmonium registration require a general explanation. Since 1843, when Alexandre Debain received a patent for the system, harmonium registration has been printed with a number or letter in a circle written above or below the staff to indicate which stops were to be drawn; a line drawn diagonally through the circle indicated that the stop was to be shut off. The stops were centered in a row above the keyboard with the *Grand jeu*, or Full Organ stop, in the middle:

0 ④ ③ ② ① G ① ② ③ ④ 0

Four sets of reeds comprised of two ranks at 8' pitch were designated ① and ④; one of 16' pitch as ②; and one of 4' as ③. Frequently a treble 16' Voix céleste was included. The 0 was a *Forte* which, when drawn, kept the expression shutters open; dynamics were controlled by levers operated by the left (bass) and right (treble) knees. The 60-note keyboard was divided at middle E so that an entirely different registration could be drawn for the left hand, playing below middle E, than for the right hand, playing middle F and above. It is important to be aware of the ② and ③, (16' and 4') as the note placement often takes advantage of the sub-octave and super-octave pitch. If ③ is indicated in the bass, it is necessary to play the passage on a 4' stop or an octave higher on an 8' stop; if ② is indicated in the treble, the right hand must play on a 16' stop or an octave lower on 8' stops.

LUDWIG VAN BEETHOVEN
PRELUDE THROUGH ALL MAJOR KEYS

In addition to three pieces composed for an organ mechanism within a clock, Beethoven left three organ works: a *Fugue in D*, written when he was thirteen, and *Zwei Präludien durch alle Dur-Tonarten*, of which he thought enough to include in his list of works as Opus 39. They were written when he was nineteen.

HECTOR BERLIOZ
THREE PIECES

These three pieces, Berlioz's only venture into organ/harmonium literature, were composed in 1845, evidently on commission by Jacob Alexandre—one of several harmonium builders in Paris during the nineteenth century—who developed and manufactured the Orgue-Mélodium. All of Berlioz's works have a literary, historical, or liturgical association except this *Toccata* and the *Rêverie et Caprice* for violin and orchestra.

The *Rustic Serenade to the Madonna* is based on themes played by Roman *pifferari*—themes reminiscent of the aria "He shall feed His flock" from Handel's *Messiah*, and the traditional Italian Christmas song "Tu schendi dalle stelle." All were inspired by the Neapolitan pipers who came down from the mountains at Christmas time and gave sacred concerts throughout Rome before statues of the Madonna. These strolling musicians were called *pifferari* after their instruments, the pifferi, a kind of primitive oboe, and the musette—a bagpipe with a reservoir made from an inflated sheepskin. Louis Spohr, in his autobiography, described the sound of the pifferi as "a coarse, powerful oboe," and the accompaniment of the bagpipe as "sounding like three clarinets together."

Berlioz, in his *Mémoires*, wrote:

> The large piffero plays the bass, the musette plays a harmony of two or three notes, over which a shorter piffero plays the melody. Then, above all that, two little, very short pifferi, played by children between the ages of twelve and fifteen, quaver trills and cadences, deluging the rustic song with a shower of bizarre ornaments. After delightfully gay refrains, loudly repeated for a long time, a slow serious piece, of quite patriarchal unctuousness, provides a stately ending to the naive symphony.

W.T. BEST
FUGUE ON A TRUMPET FANFARE

The *Fugue on a Trumpet Fanfare* was published in 1883 as a "Two-Part Fugue in C Major" in W.T. Best's *First Organ Book*. The composer thought so highly of the piece that later, in May 1894, he reworked it as a piece for two manuals with pedal as *Concert Fugue on a Trumpet*

Fanfare. The subject was slightly altered to include repeated sixteenth notes on D and G:

The performer may wish to experiment with this change.

GEORGES BIZET
THREE MUSICAL SKETCHES

Bizet won first prize in the organ class at the Paris Conservatoire in 1855. These *Trois Esquisses Musicales* were composed and published in 1858 just before Bizet left for a three-year residency as a recipient of the Prix de Rome. They are dedicated to Lefébure-Wély, then the most popular Parisian organist.

More than most of his contemporaries, Bizet was heavily influenced by the rampant "oriental-ism" of the mid-nineteenth century. While he never incorporated any authentic oriental music into his own, he frequently used descriptive titles and rhythmic and melodic figurations suggestive of the East. *Ronde Turque* is the nineteen-year-old composer's first Eastern-inspired work and clearly displays an instinct that eventually culminated in *Les Pêcheurs de perles* (1863).

Bizet's biographer, Winton Dean, recognized in the central melody of the *Sérénade* a "syncopated version of a formula that haunted Bizet all his life; its seed appears to be a phrase in the slow movement of Mozart's Clarinet Quintet."

NADIA BOULANGER
PRELUDE – PETIT CANON – IMPROVISATION

Famous as the great teacher of many post-World War I American composers, Nadia Boulanger drew students to the American Conservatory at Fontainebleau, where she taught composition and orchestration, and to her apartment in the rue Ballou, where she conducted private classes until the end of her life. Her role as teacher eclipsed her talent as an organist, though she won a first prize at the Paris Conservatoire, having been a student of Louis Vierne and Alexandre Guilmant. Boulanger composed little; included here are three of her four organ works.

JOHANNES BRAHMS
ES IST EIN' ROS ENTSPRUNGEN

Composed in June 1896, *Es ist ein' Ros entsprungen* is the eighth of Johannes Brahms's *Eleven Chorale Preludes*, Op. 122. It is based on Michael Prætorius's Christmas hymn, translated by Theodore Baker as "Lo, how a Rose e'er blooming."

ANTON BRUCKNER
ANDANTE AND POSTLUDE

The organ was Anton Bruckner's favorite instrument, and contemporary accounts describe him as one of the greatest improvisers of the second half of the nineteenth century. For eleven years (1845–56), in addition to holding the post as assistant schoolmaster at the seminary of St. Florian, near Linz, he was organist of the Augustinian abbey, where he played the second largest organ in Austria. He later won a competition for the post of organist at Linz Cathedral, remaining there for twelve years until he left to become professor of theory at the Vienna Conservatory. The *Vorspiel* and *Nachspiel* date from 1846 or 1852 and are among Bruckner's few organ works.

ERNEST CHAUSSON
ANTIPHON FOR THE MAGNIFICAT, OP. 31, NO. 6

Veni sponsa Christi is the antiphon for the Magnificat sung at First Vespers. This interlude was composed during November and December 1897 in Fiesole while Chausson was on vacation with his family. Dedicated to his daughter Annie, the *Vêpres des Vierges*, Op. 31, was published in 1898 and played for the first time by Charles Tournemire during a concert at the Schola Cantorum on March 3, 1901.

CÉSAR CUI
PRELUDE IN A-FLAT

Born the same year as Saint-Saëns, Cui was trained as an engineer and eventually became a professor at the Academy of Military Engineering. He was an authority and highly respected author on the subject of fortifications. Cui did not begin composing until 1857, though music hadalways been his avocation. He later joined Balakierev, Rimsky-Korsakov, Borodin, and Mussorgsky to form the national school of composers known as the "Russian Five," of which he was the second oldest and longest lived. Though he composed fourteen operas, Cui is remembered as a miniaturist with an uncanny ability to crystallize a particular mood. His two *Preludes* were composed for either organ or harmonium and published in 1911.

CÉSAR FRANCK
QUASI MARCIA, OP. 22

Quasi Marcia, Op. 22, was published in 1868 and dedicated to Franck's student Marie-Thérèse Miccio. It is visually deceptive, as harmonium music often is. The right hand plays on a 16' stop throughout while the left hand alternates between 8', 8' and 4', and 4' alone. Organists must be careful to equate the harmonium numbered registrations with the pitch of pipe organ stops.

EUGÈNE GIGOUT
GRAND CHŒUR DIALOGUÉ

Eugène Gigout's popular *Grand Chœur dialogué* was written (*c.* 1881) to exploit the intrinsic musical effect of two organs heard in French churches—the Grand Orgue in the tribune at the west end alternating with the orgue-de-chœur, or the small organ for accompanying the choir, in the front.

REINHOLD GLIÈRE
FUGUE ON THE THEME OF A RUSSIAN NOËL

Based on the theme of the Russian Epiphany carol "Three Kings from the Orient Came," this fugue is Glière's only organ work. It was published in 1913 soon after the composer was appointed professor of composition at the Kiev Conservatory, and follows by two years the success of his Third Symphony, *Ilya Murometz.*

CHARLES GOUNOD
INVOCATION

The *Invocation* is Gounod's own arrangement of the Offertoire from his 1855 *Messe solennelle de Sainte-Cécile,* his most famous sacred work. It was scored for orchestra but published in 1877 in arrangements by the composer for organ, and piano two- and four-hands.

ALEXANDRE GUILMANT
TROISIÈME SONATE, OP. 56

A colleague of Franck and Saint-Saëns, successor to Widor and predecessor of Gigout as professor of organ at the Paris Conservatoire, Guilmant complimented his virtuoso career with that of one of France's most influential teachers, numbering Marcel Dupré, Joseph Bonnet, and Nadia Boulanger among his pupils. He not only composed a vast quantity of organ music, but published and sold it himself. To insure wider distribution, Guilmant issued certain works in more than one version. Thus, his *Troisième Sonate,* Op. 56, completed in October 1881, appeared not only for Grand Orgue, but for harmonium or organ (Pedal ad lib.), as well. His three gifts as a composer, defined by A. Eaglefield Hull as "gifted melodist, facile harmonist, and finished constructionist," are all apparent in this Sonata, and made Guilmant one of the most popular organ composers at the turn of the twentieth century.

VINCENT D'INDY
PIECE IN E-FLAT MINOR, OP. 66

Vincent d'Indy studied organ and composition with César Franck and is remembered as his first major biographer. With Guilmant and Charles Bordes he founded the Schola Cantorum in 1896. After Guilmant's death in 1911, d'Indy appointed Louis Vierne as organ professor. In the same year he dedicated to Vierne and published the *Pièce en Mi bémol mineur,* Op. 66. Two years later the work was published on three staves as *Prélude en Mi♭ mineur.*

JACQUES-NICOLAS LEMMENS
FANFARE

Lemmens is remembered as the founder of the French school of organ playing, having taught both Guilmant and Widor. While Widor was the composer of the immortal *Toccata,* his Belgian teacher, Lemmens, pointed the way with the celebrated *Fanfare,* the first organ work to require staccato touch throughout.

FRANZ LISZT
SHEPHERDS AT THE MANGER

Since Franz Liszt had a particular interest in the reed organ—both European and American—and owned several, it is strange that he never composed for it exclusively. Rather, he published seven works for piano (harmonium *ossia*), one of which, *Shepherds at the Manger,* dating from 1874–76, combines the fourteenth-century German carol "In dulci jubilo" with a rocking cradle-song-like bass.

FELIX MENDELSSOHN
ADAGIO – ANDANTE RELIGIOSO

The Adagio of Mendelssohn's *Sonata I,* Op. 65, was composed on December 19, 1844; the Andante religioso of *Sonata IV,* on January 2, 1845. The fact that most of the movements of the six sonatas are playable on the manuals alone is evident in these two slow movements. In the last two lines of the second page of the Adagio, the low A-flat between the asterisks is to be held down by a weight, the notation making it possible to do so.

WOLFGANG AMADEUS MOZART
ANDANTE, K. 616

The *Andante fur eine Walze in eine kleine Orgel* (Andante for a cylinder in a small organ), K. 616, entered by Mozart into the catalog of his works in May 1791, was the last of three works composed for a mechanical pipe organ activated by a clockwork mechanism. The notes were pinned on a revolving drum that controlled the mechanism that activated several octaves of organ pipes. On the manuscript (the only one of the three works for mechanical clock organs that survives) Mozart originally indicated the tempo *Larghetto,* but subsequently replaced it with *Andante.*

MAX REGER
ROMANZE IN A MINOR

The *Romanze* is Max Reger's only work for harmonium. He published it in 1904 along with two alternate versions for Kunstharmonium and organ solo. Four years later, probably after the Berlin publisher Carl Simon acquired the rights, it appeared simultaneously in eleven arrangements by Richard Lange and Sigfrid Karg-Elert ranging from violin and piano to string orchestra and woodwind quartet.

GIOACCHINO ROSSINI
PRÉLUDE RELIGIEUX

Though half of the *Petite Messe solennelle* was composed as early as 1831, the completed work was not performed until March 14, 1864, when it was sung at the home of the banker Pillet-Will. Scored for four soloists, chorus, two pianos, and harmonium, Gioacchino Rossini later orchestrated it. The *Prélude religieux* is the Offertoire of the Mass and includes a sixteen-measure piano introduction to the present extended organ solo, the last four measures of which are again indicated for piano. These latter measures can either be played on the organ or omitted, the organ solo concluding with an F-sharp major chord.

CAMILLE SAINT-SAËNS
THREE PIECES, OP. 1

With Berlioz and Alkan, Saint-Saëns was one of the first important composers to write for the harmonium, the free-reed instrument with its single manual divided between middle E and F, permitting contrasting registrations between the left and right hands.

In *Méditation*, Saint-Saëns exploits the instrument's full dynamic range from one stop to full organ (or Grand Jeu on the harmonium, which brings on 16', 8', and 4' stops) with the expression swells open and closed.

Barcarolle is directly inspired by Mendelssohn's "Venetian Boat Song" (*Songs Without Words*, Op. 62,

No. 29), composed about ten years before Saint-Saëns's work. The right hand plays an octave higher on a 16' register throughout the piece, thus keeping the theme distinct from the accompaniment, and skillfully avoiding crossing over the manual division between middle E and F.

The *Prière* was a genre popular with nineteenth-century organ and piano composers. César Franck's *Prière* is the incontestable masterpiece of the genre but Saint-Saëns's is noteworthy for its unusual 11/4 meter in which each measure is subdivided into two groups of four and one of three beats.

Composed in 1852, the *Trois Morceaux* were published in the fall of 1858.

LOUIS VIERNE
COMMUNION, OP. 8

The great organist of Notre-Dame Cathedral, Louis Vierne, has left a significant body of superb pieces for manuals only; his *24 Pièces en style libre*, published in 1914, remain masterpieces of the genre. This beautiful *Communion*, with a theme reminiscent of the third-act quintet from Wagner's *Die Meistersinger*, was composed around 1896 and premiered by the composer in a recital given on the Merklin organ at Saint-Jean-de Malte, Aix-en-Provence, in January 1897. The present two-stave version includes harmonium registration as well as organ registration and manual changes.

CHARLES-MARIE WIDOR
ANDANTE SOSTENUTO

Charles-Marie Widor's *3me Symphonie* for organ and orchestra, Op. 69, was written for the inauguration of Victoria Hall, in Geneva, Switzerland. The composer conducted it at the opening concert on November 28, 1894. The Symphony is in two movements; this *Andante sostenuto*, which ends the first movement, provides the slow movement.

Rollin Smith

Prelude Through All Major Keys, Op. 39, No. 2

Praeludium in allen Dur-Tonarten

Ludwig van Beethoven
1779–1827

Andante, K. 616

Wolfgang Amadeus Mozart
1756–1791

THREE PIECES

I. Rustic Serenade to the Madonna
Sérénade agreste à la Madone

Hector Berlioz
1803–1869

Allegro assai.

II. Toccata

Hector Berlioz

III. Hymn for the Elevation

Hymne pour l'élévation

Hector Berlioz

Andante
Vorspiel

Anton Bruckner
1824–1896

Postlude
Nachspiel

Anton Bruckner

Full Organ

Adagio from Sonata I, Op. 65

Felix Mendelssohn
1809–1847

Andante religioso from Sonata IV, Op. 65

Felix Mendelssohn

THREE PIECES, OP. 1
I. Méditation

Camille Saint-Saëns
1835–1921

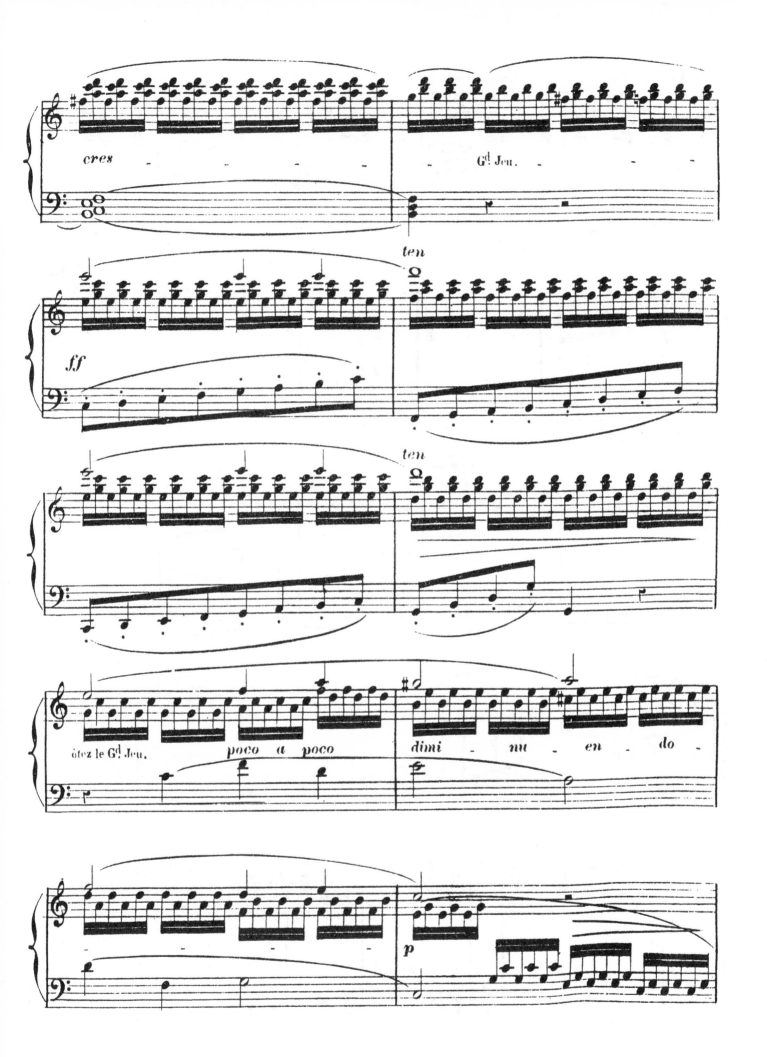

II. Barcarolle

Camille Saint-Saëns

III. Prière

Camille Saint-Saëns

Quasi Marcia, Op. 22

César Franck
1822–1890

THREE MUSICAL SKETCHES
I. Turkish Rondo
Ronde Turque

Georges Bizet
1838–1875

otez (1) mettez (2)

(1)

dim.

II. Sérénade

Georges Bizet

III. Caprice

Georges Bizet

Fanfare

Jacques-Nicolas Lemmens
1823–1881

Prélude religieux

from *Petite Messe Solennelle*

Gioacchino Rossini
1792–1868

Andantino mosso.

Invocation

Charles Gounod
1818–1893

The Shepherds at the Manger
"In dulci jubilo"

Franz Liszt
1811–1886

Allegretto pastorale.

Grand Chœur dialogué

Eugène Gigout
1844–1925

Allegro moderato quasi maestoso ♩ = 69

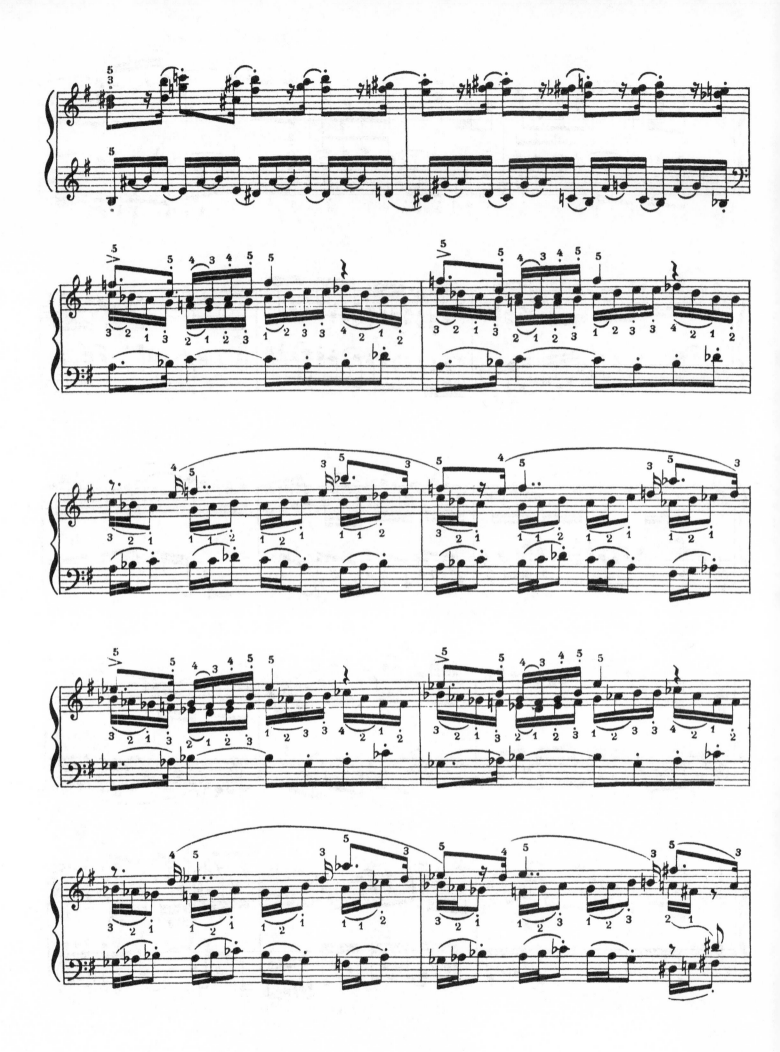

Third Sonata in C Minor, Op. 56

I. Preludio

Alexandre Guilmant
1837–1911

II.

Molto Adagio. ($\flat = 40$.)

(*) Cet Adagio peut aussi s'exécuter sur l'harmonium avec
(céleste) en jouant le tout une octave plus haut.

III. Fuga

Communion, Op. 8

Louis Vierne
1870-1937

Andante sostenuto

from the Third Symphony for Organ and Orchestra, Opus 69

Charles-Marie Widor
1844-1937

Fugue on a Trumpet Fanfare

W.T. Best
1826–1897

A Rose Breaks Into Bloom, Op. 122, No. 8

Es ist ein' Ros' entsprungen

Johannes Brahms
1833–1897

(Man. II.)

Antiphon for the Magnificat, Op. 31, No. 6

Ernest Chausson
1855–1899

Piece in E-flat Minor, Op. 66

Vincent d'Indy
1851–1931

Prélude

Nadia Boulanger
1887–1979

Petit Canon

Improvisation

Nadia Boulanger

Prelude in A-flat

César Cui
1835-1918

Fugue on the Theme of a Russian Noël

Reinhold Glière
1875–1956

Romanze in A Minor

Max Reger
1873–1916